I0814059

BEAUTIFUL BIOMES

TUNDRA BIOME

by Elizabeth Andrews

Cody Koala

An Imprint of Pop!

popbooksonline.com

abdobooks.com
Published by Pop!, a division of ABDO, PO Box 398166, Minneapolis, Minnesota 55439.

Printed in the United States of America, North Mankato, Minnesota

102021
012022

THIS BOOK CONTAINS RECYCLED MATERIALS

Cover Photo: Shutterstock Images
Interior Photos: Shutterstock Images, 1, 5 (top) (bottom left) (bottom center), 7, 8, 9, 13, 14–15, 17 (top) (bottom left) (bottom center), 18; AscentXmedia/Getty Images, 10–11; Bosca78/Getty Images, 20–21

Editor: Tyler Gieseke
Series Designer: Laura Graphenteen

Library of Congress Control Number: 2021942295

Publisher's Cataloging-in-Publication Data
Names: Andrews, Elizabeth, author.
Title: Tundra biome / by Elizabeth Andrews
Description: Minneapolis, Minnesota : Pop!, 2022 | Series: Beautiful biomes | Includes online resources and index.
Identifiers: ISBN 9781098241056 (lib. bdg.) | ISBN 9781098241759 (ebook)
Subjects: LCSH: Tundras--Juvenile literature. | Biotic communities--Juvenile literature. | Habitats--Juvenile literature. | Life zones--Juvenile literature. | Tundra animals--Juvenile literature. | Tundra plants--Juvenile literature. | Tundra ecology--Juvenile literature.
Classification: DDC 577.5--dc23

Hello! My name is

Cody Koala

Pop open this book and you'll find QR codes like this one, loaded with information, so you can learn even more!

Scan this code* and others like it while you read, or visit the website below to make this book pop.

popbooksonline.com/tundra-biome

*Scanning QR codes requires a web-enabled smart device with a QR code reader app and a camera.

Table of Contents

Chapter 1

The Coldest Biome

A biome is a large, natural area. It is known for the plants and animals that live there, and its **climate**.

tundra

forest

freshwater

Watch a video here!

Chapter 2

Tundra Biomes

The tundra biome is usually a cold and icy place. There are two kinds of tundra. Arctic tundras are far north in the **Arctic Circle**. The winters there are very long. **Temperatures** stay around -30°F (-34.4°C).

Learn more here!

Summers in the Arctic are short. They can get up to 50°F (10°C). All plant growth happens during the short

summer months. Animal populations grow too. That's when babies are born.

Alpine tundras are found high in the mountains. They can occur in any mountain range that gets tall enough.

It is cold and windy up there. Temperatures are usually 10°F (-12.2°C) in the winter and 50°F (10°C) in the summer.

Temperatures go down when air pressure decreases. Air pressure lowers high up in the mountains, so it gets colder there.

Chapter 3

Tundra Plants

It is hard for plants to grow in tundras. The **climate** is too cold. Wind blows too hard for trees to survive. The ground never **thaws** enough to support large plants with deep roots.

NO TREES IN THE TUNDRA
tundra
treeline
below treeline
Learn more here!

Most plants in tundras grow close to the ground. Shrubs, mosses, grasses, and flowers have short roots.

The summer sun shines all day and night in the Arctic tundra. This helps plants grow quickly.

They grow quickly in the summer. Flowers bloom for only a few days.

Chapter 4

Tundra Animals

Tundra animals have many **adaptions**. Some animals' fur changes color with the seasons. In winter they are white like the snow. In summer they are darker like the ground.

Complete an activity here!

Arctic ground squirrels survive the cold winter by **hibernating**. Eight months can feel like 12 days to them during hibernation.

There are tundra animals of all shapes and sizes. Small rodents like Arctic ground squirrels and lemmings burrow in the ground. Giant polar bears live near the ocean. They hunt seals beneath the water.

Alpine tundras have different kinds of animals. These animals are built to live high up in the mountains. Musk ox and mountain goats have hooves. They are good climbers.

Making Connections

Text-to-Self

Which tundra animal do you think has the best adaptions for the cold climate?

Text-to-Text

Have you read a book about any other biomes? If so, how are they similar to and different from tundras?

Text-to-World

There are groups of people that live in the tundra. Do you think life is easy for them? What challenges would they face?

Glossary

adaption – a change in a plant or animal that helps it survive in a certain place.

Arctic Circle – an imaginary line drawn around the top of the earth that marks the boundary of the Arctic.

climate – weather conditions that are usual in an area over a long period of time.

hibernate – to spend a period of time, such as the winter, in deep sleep.

temperature – a measure of hotness or coldness.

thaw – to go from being a frozen solid to a soft solid.

Index

Online Resources

popbooksonline.com

Thanks for reading this Cody Koala book!

Scan this code* and others like it in this book, or visit the website below to make this book pop!

popbooksonline.com/tundra-biome

*Scanning QR codes requires a web-enabled smart device with a QR code reader app and a camera.